YOUR KNOWLEDGE HAS VALUE

Purvesh Shah

Synthesis of some complex molecules

GRIN Verlag

Bibliografische Information der Deutschen Nationalbibliothek:

Die Deutsche Bibliothek verzeichnet diese Publikation in der Deutschen National-
bibliografie; detaillierte bibliografische Daten sind im Internet über http://dnb.d-
nb.de/ abrufbar.

Imprint:

Copyright © 2014 GRIN Verlag GmbH
Druck und Bindung: Books on Demand GmbH, Norderstedt Germany
ISBN: 978-3-656-61351-0

This book at GRIN:

http://www.grin.com/en/e-book/270173/synthesis-of-some-complex-molecules

SYNTHESIS OF SOME COMPLEX MOLECULES

Dr. Purvesh J. Shah

Department of Chemistry,

Shree P.M.Patel Institute of P.G.Studies and research in science, Anand-

388001, India

Affilited to Sardar Patel University, Vallabh Vidyanagar 388 120, India

Contents

Mesoporphyrin IX

Mesoporphyrins:

- They are degradative products of Haemin.

- If Iron is removed &vinyl group is reduced to Ethyl, then 15 Mesoporphyrins with 4 pyrrole moiety are possible.

- One of them is Mesoporphyrins IX.

- This is identical with Mesoporphyrin obtained from Haemin.

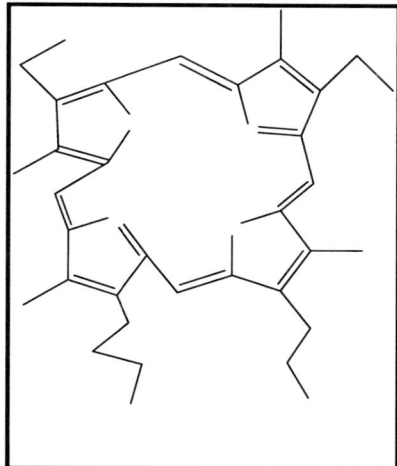

Haemin-

- Protin free fragment.

- Responsible for the colour of Blood.

- Involves in Oxygen transfer.

- Belongs to general class of compounds known as "Porphyrine".

- "Hans Fischer" carried out a very extensive work & established its structure.

- Large no. of degradative products was studied.

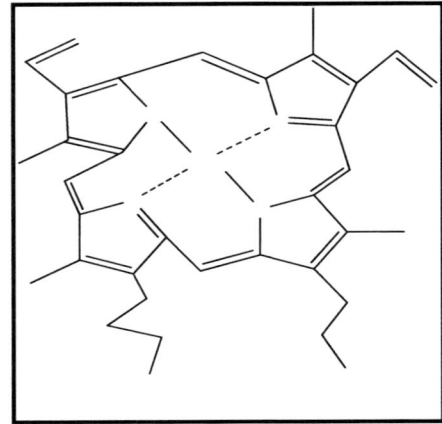

To synthesized Mesoporphyrins:- pyrrole derivatives synthesized.

Pyrrol (3) + Pyrrole (4) $\longrightarrow$ A & B ring

Pyrrole (8) $\longrightarrow$ C & D ring.

*Synthesis of substituted pyrrole:-

General reaction-

*Synthesis of Pyrrole (3):

*Synthesis of Pyrrole (4):

*Synthesis of Pyrrole (8):

Mechanism:

5

*Preparation of A+B:

*Pyrrole (8) +pyrrole (8)--------ring C+D

*Mechanism:- Step 8 to Step12

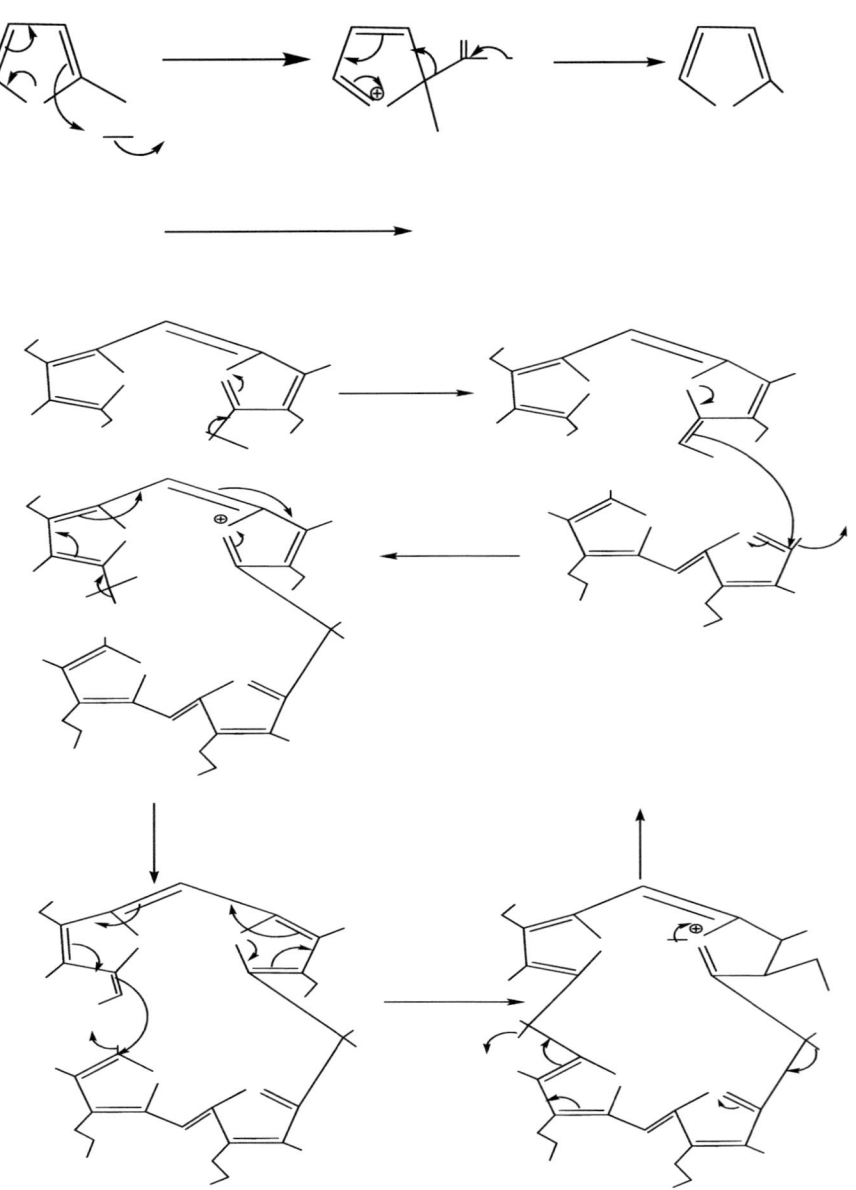

*Cephalosporin C:

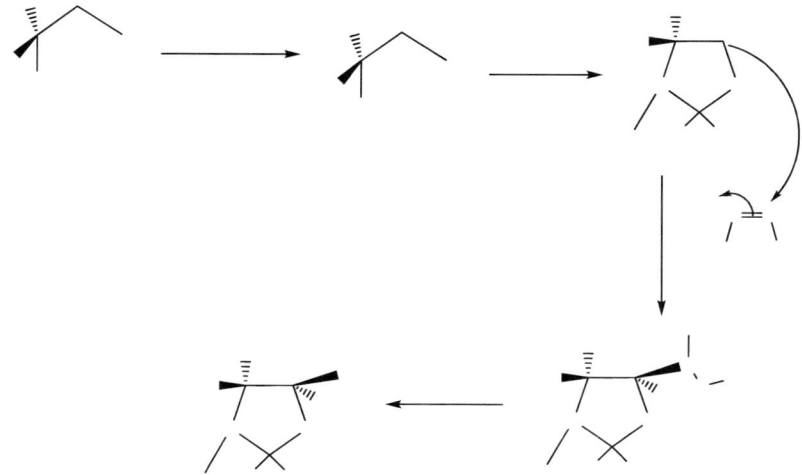

- Like penicillin's, the Cephalosporin C are antibiotics with a wild range of activity.

- Cephalosporin C is relatively inactive compared to other related compounds.

- Penicillin's resistant organisms remain active against Cephalosporin C.

- Cephalosporin C is yet not much used.

- Second line of defense.

- R.B. Woodward & Co-workers, J. American Soc.,88(1966)852.

Mechanism: 4 to 5:

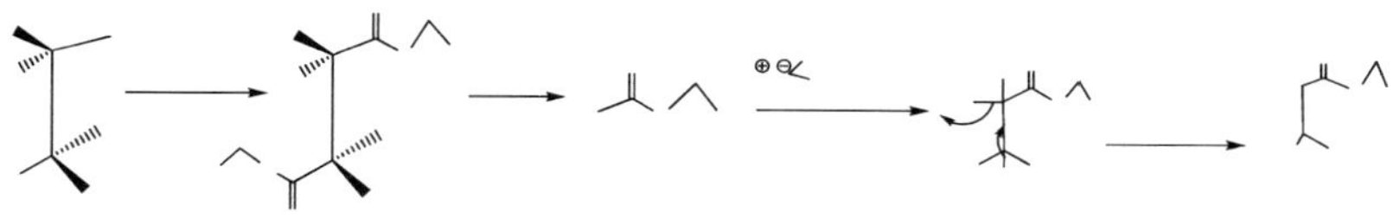

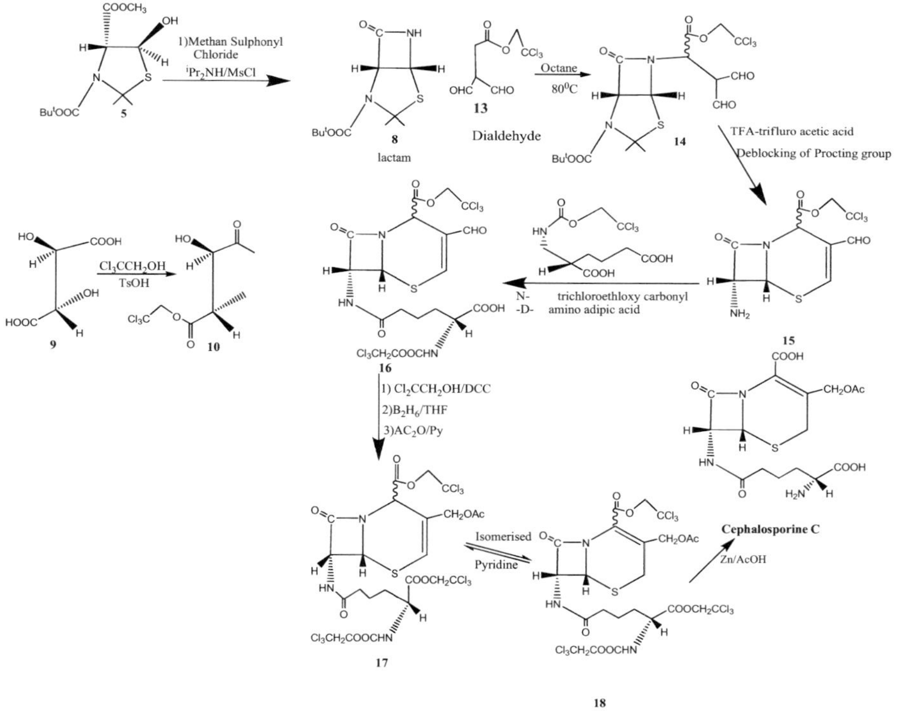
COOCH3
OH
1)Methan Sulphonyl Chloride
iPr2NH/MsCl
5
8
lactam
13
Dialdehyde
Octane
80°C
14
TFA-trifluro acetic acid
Deblocking of Procting group
15
N-trichloroethloxy carbonyl -D- amino adipic acid
Cl3CCH2OH
TsOH
9
10
16
1) Cl2CCH2OH/DCC
2)B2H6/THF
3)AC2O/Py
17
Isomerised
Pyridine
18
Zn/AcOH
Cephalosporine C

18 more stable than 17
Conjugation

Coenzyme A

- Complex thio derivative.

- Contains functional group –SH.

- -SCOCH3 of Coenzyme A acts as a carrier of acyl group In biochemical system.

- Synthesized by J.G.Moffatt & H.G.Khorana,1959.

- J.American Chemical Society,83(1961)663.

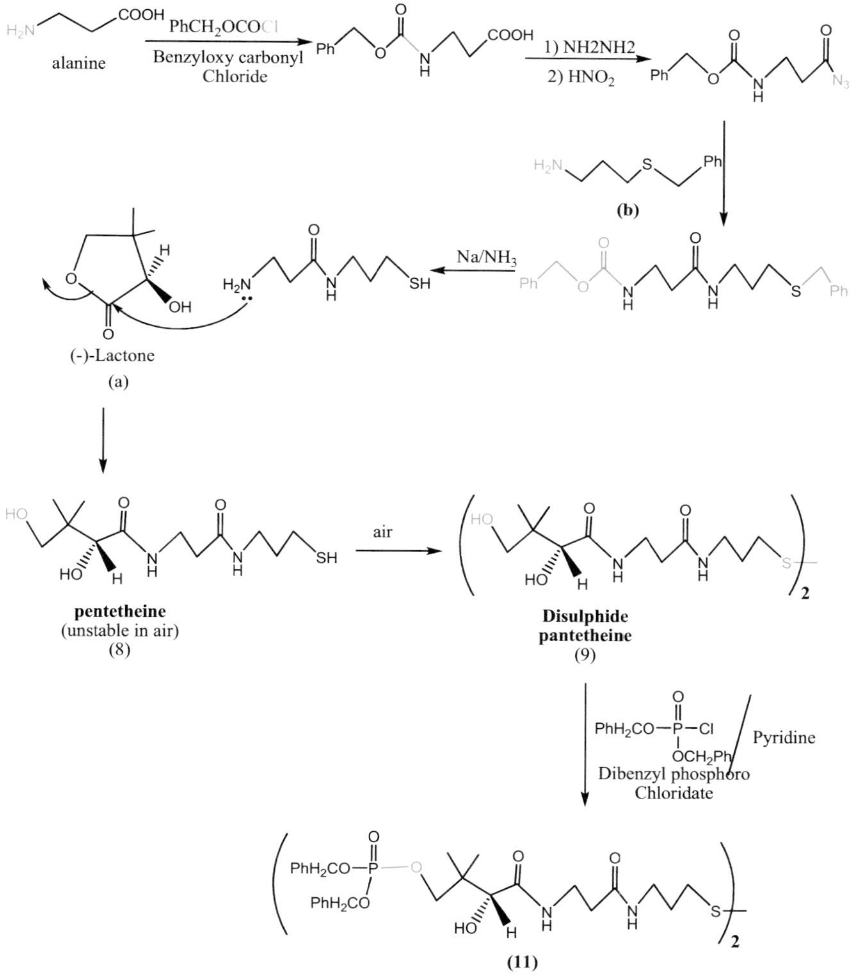

alanine

PhCH₂OCOCl
Benzyloxy carbonyl
Chloride

1) NH2NH2
2) HNO₂

(b)

H₂N—S—Ph

Na/NH₃

(-)-Lactone
(a)

pentetheine
(unstable in air)
(8)

air

Disulphide
pantetheine
(9)

PhH₂CO—P—Cl
OCH₂Ph
Dibenzyl phosphoro
Chloridate

Pyridine

PhH₂CO
PhH₂CO

(11)

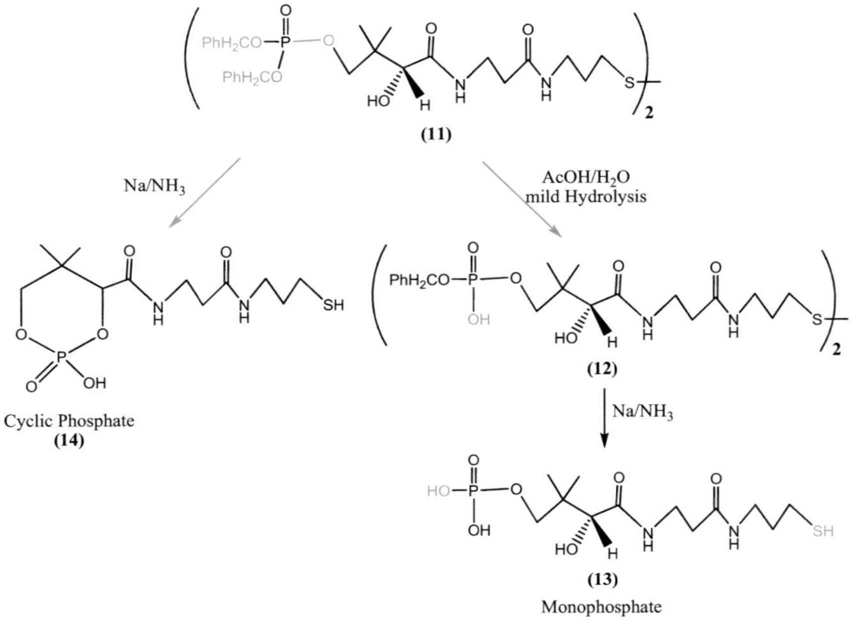

(11)

Na/NH₃

AcOH/H₂O
mild Hydrolysis

Cyclic Phosphate
(14)

(12)

Na/NH₃

(13)
Monophosphate

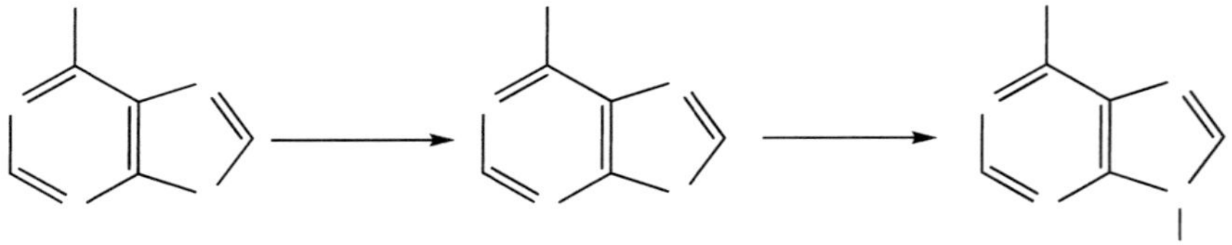

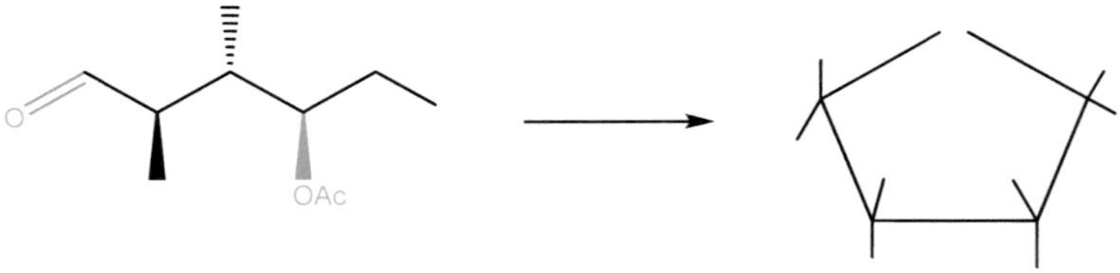

HgCl

NHAc
(c)

H

Br

OAc AcO

O

H

H

CH₂OAc

H

H

Bromo sugar deri.
(D)

H

O

H

N

N

N

N

OAcAcO

H

H

CH₂OA
c

NHAc

HCl/MeOH

H

O

H

N

N

N

N

OH HO

H

H

CH₂OH

NH₂

Adenosine

PhCH₂O—P—Cl $\quad$ / Pyridine

O

OCH₂Ph

PhH₂CO

H

O

H

Ad

O

P

O

H

H

CH₂OR

(G)

+

H

O

H

Ad

OH RO

H

H

CH₂OR

(F)

+

H

O

H

Ad

OR HO

H

H

CH₂OR

(E)

Where, Ad= Adenine $\quad$ R= —P—OCH₂Ph

O

OCH₂Ph

R'= —P—OH

O

OH

E + F $\quad \dfrac{\text{1) H}_3\text{O}^+}{\text{2) H}_2/\text{Pd}}$

H

O

H

Ad

OH R'O

H

H

CH₂OR'

(F)

+

H

O

H

Ad

OR' HO

H

H

CH₂OR'

(E)

1) DCC dicyclohexyl carbodiimide

2) Morphine HN O

15

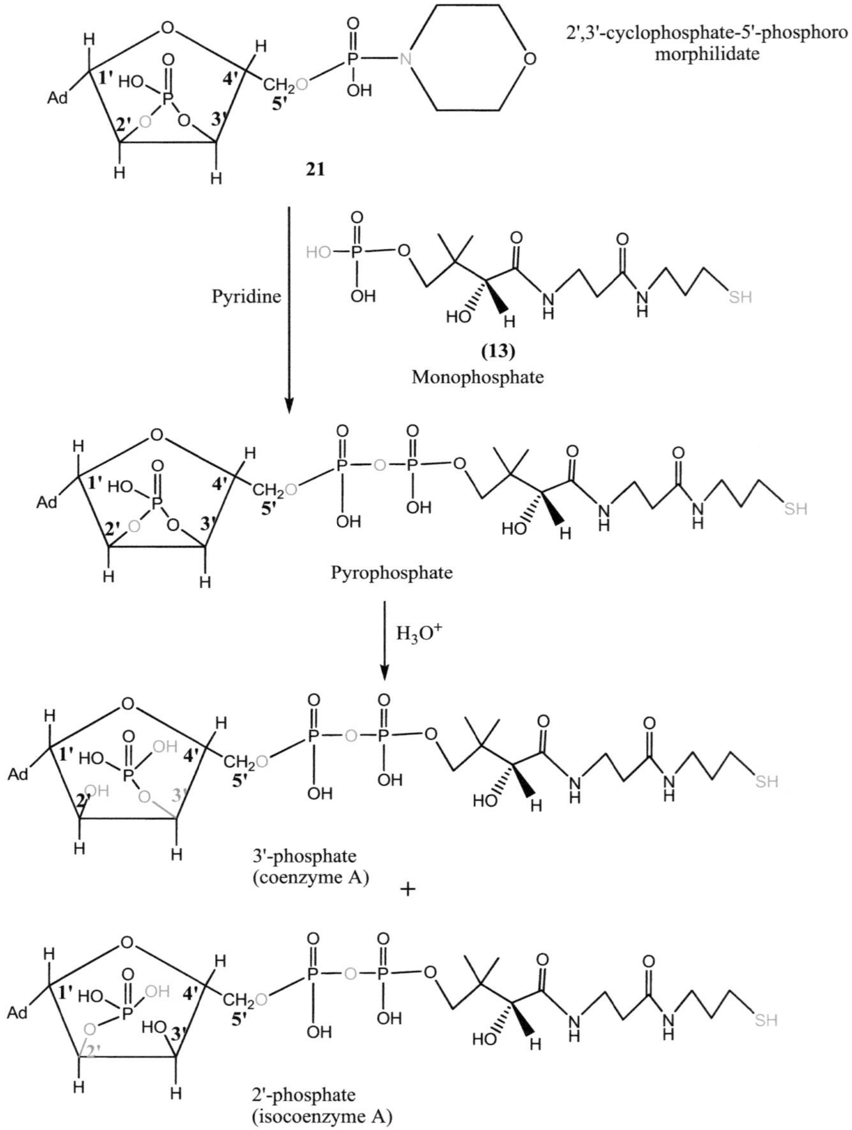

2',3'-cyclophosphate-5'-phosphoro morphilidate

21

Pyridine

(13)
Monophosphate

Pyrophosphate

H_3O^+

3'-phosphate
(coenzyme A)

+

2'-phosphate
(isocoenzyme A)